Jorge Alfredo González Pérez
Augusto Manuel Kahali Petuleinge

Development of actions in the process of agricultural marketing of beans

Jorge Alfredo González Pérez
Augusto Manuel Kahali Petuleinge

Development of actions in the process of agricultural marketing of beans

Development of actions to improve the process of agricultural commercialization of bean (Phaseolus vulgaris L.) in the markets

ScienciaScripts

Cover image: www.ingimage.com

This book is a translation from the original published under ISBN 978-613-9-72774-2.

Publisher:
Sciencia Scripts
is a trademark of
Dodo Books Indian Ocean Ltd. and OmniScriptum S.R.L publishing group

120 High Road, East Finchley, London, N2 9ED, United Kingdom
Str. Armeneasca 28/1, office 1, Chisinau MD-2012, Republic of Moldova, Europe
Managing Directors: Ieva Konstantinova, Victoria Ursu
info@omniscriptum.com

Printed at: see last page
ISBN: 978-620-8-37321-4

DEDICATORY

- To my family, in particular my parents, for everything they have done for my life in every way.
- To my siblings, for whom I have a lot of love and appreciation.
- I would then like to dedicate it to my friends at the sports activity group called "The Friends", for their invaluable support throughout this odyssey.
- To my colleagues on the Agricultural Engineering degree course at the Ondjiva Polytechnic Institute.
- To the entire scientific community, as this work will serve as a source of knowledge, especially for those in the field of Agricultural Engineering.

ACKNOWLEDGMENTS.

- I would like to thank GOD, the almighty creator of heaven and earth, for the health, patience and wisdom he has given me in carrying out this work and, above all, for the life he has given us so far.
- MANDUME YANDEMUFAYO UNIVERSITY, IPO, for giving me the privilege and honor of a very fruitful education.
- To my excellent teachers, administrative staff. cleaners, custodians and other staff at this Higher Education Institution, who directly and indirectly contributed to my education.
- I would like to thank my classmates and fellow students for the great and unforgettable times we had together.
- I am very grateful to my eternal colleague "Manuel Francisco", may God have him in all his glory and may the moments spent in his company never be forgotten, for they will be forever in my memory.

SUMMARY

With the aim of proposing a group of actions to improve the bean marketing process, this study was carried out in the informal markets of Oshomokyo and Germany, between February and May 2022. This study was considered to be exploratory-descriptive because its characteristics allowed it to describe a population, learn about consumer behavior and qualities in relation to products, and learn about supply and demand trends. A non-experimental design was used, with the use of scientific observation and research. The field information was collected in informal markets where 8 randomly selected consumers (customers) and 11 vendors were interviewed, 6 in Germany and 5 in Oshomokyo. The information collected, was processed using tables and graphs in the Microsoft Word and Excel for Windows 10 program. According to the results: The marketing of beans works under the sales model: producer-acopiator or intermediary-wholesaler and consumer. In the Germany and Oshomokuyo markets, 92% of the consumers surveyed said that the beans on offer are good, but they are not sold at different prices, and Oshomokuyo is where the greatest variety of beans is sold. The situations found in the SWOT in the two places presuppose similar reactions, but it should be noted that a weakness or threat in Germany will not necessarily be the case in Oshomokuyo. The actions set out must be carried out in different ways, but the most efficient way is through government action projects with a clear intention in their execution.

Key words: beans, action plan, informal markets

Table of contents

I. INTRODUCTION

Beans *(Phaseolus vulgaris* L.) have been cultivated by humans for thousands of years. Its origins lie on the American continent, with the oldest archaeological evidence dating back some 8,000 to 10,000 years in South America (Kaplan *et al.* 2016). As for its domestication, it is argued that there are differences.

Worldwide, beans (considering all species) are grown in more than a hundred countries. Brazil and India are the largest producers and together account for more than 35% of world production. Around 23 million tons of beans are produced annually, covering an area of approximately 30 million hectares (FAO, 2020).

The volume of beans traded on the international market is considered low (FAO, 2019). Considering that only 14% of world production is destined for export. One of the reasons given for the low international trade is the wide variety of bean types and the differences in eating habits between consumer countries.

The "marketing of agricultural products" is the entire process that goes from the moment the product leaves the company, the farm or the holding of the producer or entrepreneur to the hands of the consumer, Fisher and Navarro (2018). The same authors state that marketing not only refers to the action of buying or selling, i.e. the change of ownership of the good, but also to the physical aspects of transportation (change of place), storage (change in time), conditioning and processing (change of form).

Marketing is the main problem faced by the agricultural sector, since there are no companies dedicated to this activity, while the number of intermediaries is increasing, creating an inequality between direct prices to producers and consumers. Finally, this makes it necessary for small producers to find new alternatives to market their products, in order to form leading companies dedicated to marketing agricultural products at fair prices and achieve the integration of different producers to improve the quality, quantity and prices of products on the market, (NEVES E CASTRO, 2017).

In Angola, beans are grown by small and medium-sized producers, in a variety of production systems (including fully mechanized, from planting to harvesting) and in all regions of the country, ranging from production for the subsistence of the farmers themselves, to large producers at corporate level, with a considerable technological investment. This difference occurs mainly because the crop can be produced in any region of the country, regardless of the technological level, and also because consumption is widespread throughout the country, in both rural and urban regions. (Jornal de Angola, 2021).

According to Moreira (2021), Angola has an annual deficit of beans and in the 2019/2020 period, Angola stood at just 634,205 tons, a negligible percentage of world production of 26.84 million tons per year. In the same period, 344,762 tons of beans were destined for human consumption in 2019/2020, while 46,580 for seeds and 59,691 for other uses, with losses estimated at 13,111 tons.

Bean cultivation is dominated by the family farming sector, which employs 2,936,198 families in a total area of 5,325,952 hectares per year, while the business sector brings together a total of 8,826 small, medium and large farms and a total area of 493,990 hectares per year (Jornal de Angola, 2021).

According to the same source, "the cultivation of beans is one of the most profitable, because it is a product that is part of eating habits almost everywhere in the country. Its production is concentrated in the provinces of Bié, Huambo, Malanje, Uíge, Cuanza-Sul, Huíla and Benguela, but all the provinces have potential". It is produced during three seasons of the year, but you can also bet on greenhouse cultivation, which allows you to have harvests at any time of the year.

In Cunene province, specifically in Ondjiva commune, beans are an important part of the local diet, so marketing levels are variable and present their own difficulties. Unfortunately, at the moment there is not enough information to know how this product performs in informal markets.

Based on the above elements, the following **scientific problem** was formulated**:** How is the agricultural marketing of beans *(Phaseolus vulgaris* L.) carried out in the informal markets of Ondjiva Commune in the Cuanhama Municipality?

Based on the scientific research problem, the following **hypothesis** is basically defended**:** The efficiency of the agricultural marketing of beans *(Phaseolus vulgaris* L.) *could* be improved if the factors affecting this process in the informal markets of Ondjiva Commune in the Cuanhama Municipality were known.

To answer the problem raised, the following **General Objective** was defined:

Designing actions to improve the agricultural marketing process for beans *(Phaseolus vulgaris* L.) in the informal markets of Ondjiva Commune in Cuanhama Municipality.

In order to achieve the general objective, the following **specific objectives** were formulated:

- ✓ To characterize the development of the agricultural marketing process for beans *(Phaseolus vulgaris* L.) in the informal markets of the Ondjiva Commune in the Cuanhama Municipality.
- ✓ To identify the factors that affect the process of agricultural marketing of beans (*Phaseolus vulgaris* L.) in the informal markets of Ondjiva Commune in Cuanhama Municipality.
- ✓ Proposing a group of actions to improve the process of marketing beans (*Phaseolus vulgaris* L.) in the informal markets of Ondjiva Commune in the Cuanhama Municipality.

II. LITERATURE REVIEW

2.1 Taxonomy of the common bean crop *(Phaseolus vulgaris* L.)

The common fenugreek belongs to the genus *Phaseolus* and its taxonomic location is: Rainho: Plantae; division: Magnoliophyta; class: Magnoliopsida; subclass: *Rosidae;* order: *Fabales;* family: *Fabaceae*; genus: *Phaseolus* and species: *Phaseolus vulgaris* L.

2.2 Overview of the common bean crop *(Phaseolus vulgaris* L.)

The origin of *Phaseolus Vulgaris* has been a hotly debated topic among historians, with many believing that this crop originated in Europe and others pointing out that it originated in certain regions of America, from where it was distributed to other continents. After more detailed studies of its origin, there have been sporadic appearances in American countries such as Peru, Mexico, Guatemala, Chile and Bolivia, and the conclusion has been reached that it is of American origin. This certainty is based on data obtained from 1500 isolated points that appear in different descriptions and references (González, 2018).

Beans are one of the most important food legumes for human consumption. Its production covers diverse agro-ecological areas. This legume is grown virtually all over the world (FIEP, 2017).

2.3 Nutritional and economic importance of the crop

The common bean known as Feijão, Feijão de rimón, Feijão, Poroto and Caraota is a crop of great importance in human nutrition due to its high nutrient content. In Latin America it is an essential component of the diet as it is an important source of protein (Socorro and Martín, 2009).

According to Castiñeiras (2011), beans are the legume that has been the subject of the most studies in Latin America, as they are the main source of protein and form an important part of people's eating habits. Scientific advances corroborate the need to incorporate and maintain this food in the conventional diet, due to its proven nutritional and medicinal values. Likewise, its contribution to protein stands out, so that in food guides it is part of the group of foods characterized by their protein content.

Beans provide vitamins such as the B-complex and folic acid. Legumes are rich in dietary fiber and minerals such as calcium, iron, copper, zinc, phosphorus, potassium and magnesium. Beans are one of the main sources of soluble fiber in the common diet and help reduce cholesterol (Silva, 2018).

2.4 Morphology of the common bean *(Phaseolus vulgaris* L.)

Beans are herbaceous plants, with a relatively short biological cycle and an annual character, varying in size and habit, as there are varieties of determinate and indeterminate growth (small shrubs and climbers) as described (Socorro a Martin, 2009).

In its morphological constitution, the bean is made up of the following organs:

Root: the root system is made up of a main root, as well as a large number of secondary roots. When it germinates, it is fast-growing, its active layer developing between 0.20 - 0.40 cm deep and within a radius of 0.15-0.30 cm. With numerous lateral branches, this crop has the ability to fix atmospheric nitrogen through symbiosis with bacteria of the genus Rhizobium from the formation of nodules on its roots, Gepts and Debouck (2015).

Stem: The stem is made up of nodes and internodes that vary in size, and from each node a leaf emerges. Its height depends on its growth habit (determinate or indeterminate). They are called determinate when they reach a low height (0.20-0.60 cm) and have an inflorescence at their end, while the indeterminate ones can reach two to ten meters in length and have no inflorescence at their terminal bud (Ferreira *et al.* 2012).

Leaves: the leaves are alternate, made up of three leaflets (two lateral and one terminal or central). The leaflets are large, oval with acuminate ends and have a pointed shape (Unifeijão, 2011).

According to Wander (2007), "The leaflets are oval or rhomboid in shape. The leaves are alternate, trifoliate and dark or light green in color. The leaflets vary in shape: oval, deltoid and cuneiform. There are also trifoliate leaves.

Inflorescence: can be terminal (these are only present in determinate growth varieties) and axillary, which are present in both growth habits. The flowers have five unequal petals: a standard, two fused petals that form the keel and two "wings". The flower is symmetrical and can be of various colors: white, pink, yellow, violet (Diniz, 2018).

Fruit: this is a vegetable usually known as a pod, with an enlarged shape, which can have different colors such as cream, coffee or purple. The pod contains between three and nine seeds, but the normal number is between five and seven. They are reniform in shape, although they can also be round, ovoid, elliptical, small, almost square, and ovoid (Muñoz and Singh, 2003).

Seed: depending on the color, you can find grains with a uniform color, for example black, red and white, you can also find two colors with different variations within the group, and finally even three different colors. The state of physiological maturity, or the end of the grain's growth, is reached when it has an average moisture content of 52 to 54%. The color of the grains is green at the beginning of their growth, until they reach a humidity slightly above or very close to 60%; thereafter the grains gradually acquire the characteristic colors of each cultivar, to obtain their definitive color at the state of physiological maturity (Diniz, 2018).

The seeds of this crop have the property of quickly losing their moisture once they are ripe, and can be stored without major difficulties, as their integuments are very impermeable, although their thickness is a characteristic that depends on the variety and type of bean (Socorro and Martín 2009).

2.5 Marketing alternatives or strategies

An "alternative" is defined as a procedure, mechanism, method or option through which a producer can sell or influence the terms of sale of their product. The main alternatives available to a producer are: cash sale at harvest time, pre-harvest sales contract, staking for speculation, authorized price, price to be fixed, average price or joint sales (Mendes *et al.,* 2007).

2.5.1 Strategy principles

Agricultural production, as an economic activity and associated with a multitude of variables that condition both technological results and profitability, also requires the adoption of a strategy, those of a technical nature (physical and biological aspects) and those of an institutional and human nature, considered exogenous or endogenous to the farm (Neves, 2010). The author also points out that the agricultural marketing strategy is the set of principles, objectives, actions and priorities for the development of agricultural trade, based on private initiative, market forces and the regulatory and facilitating role of the state, and must comply with the following principles:

- Compliance with the fundamental options;
- Definition of commerce as an activity essentially based on private initiative and the necessary link between production and consumption;

- The need to promote and modernize the commercial network and related services;
- The need to promote the marketing process taking into account economic interests, particularly in the protection of agricultural production and trade.

2.6 Proposed actions. Definition

A pattern or plan that integrates an organization's main goals, policies and sequence of tasks into a coherent whole. The authors relate a well-formulated set of actions to a singular and viable stance achieved by its own competencies, but controlling its internal deficiencies and arrangements made by intelligent opponents (Senaes, 2013).

2.7 Generalities about the marketing process

Agricultural marketing plays an extremely important role in the national economy, being the main, if not the only, source of income for the population in rural areas, where the majority of people live off subsistence farming. Marketing is one of the driving factors behind the links between the producer/peasant and the market, in terms of the economic relationship between rural and urban areas (Agronegocios, 2014).

For Barros and Martines (2017), agricultural marketing refers to a set of functions or activities of transformation and addition of utility through which goods and services are transferred from producers to consumers. Thus, in the broadest definitions, the marketing of goods is also a production process. In addition, marketing includes activities that result in the transformation of goods, through the use of productive resources - capital and labor - that act on agricultural raw materials.

Marketing is based on the transition of goods to a final destination, whenever standards and procedures are complied with and customer satisfaction is obtained, specifically in agricultural products, which have a more demanding market, where the quality of the products must stand out, in order to obtain quality sales for the company and customer satisfaction (Piza and Welsh, 2018).

Since ancient times, man has had to rely on exchange as the first form of commerce in order to survive and then develop. In semantics, commercialization is a synonym for marketing in English or marketing in Spanish, the commercial term meaning to negotiate by buying and selling something (Pinho *et al.,* 2014).

The texts on agricultural marketing present various definitions with varying degrees of breadth. In a narrower sense, marketing has been defined as the set of operations that begin with the creation of agricultural products and end with their use by the final consumer. In this context, marketing means the continuation of the production process (Porter, 2006).

Agricultural marketing is not just the sale of produce, it is a continuous and organized process of moving agricultural production along a marketing channel or system, where the product undergoes transformations, differentiations and value additions. The facilities (utilities) that agricultural products undergo are of possession, form, time and place, thus adapting them to the tastes and preferences of end consumers (Neves, 2010).

Agricultural marketing is the most complex activity in the agricultural system, as it is the moment when production becomes a commodity. This condition reflects the dynamic that market integration, comprising various segments and sectors, almost appropriates production and begins to impose quantity and quality targets, forming chains, networks or production arrangements (Joseph, 2015).

2.8 Terms used in the marketing process

According to Navolar *et al.* (2010) the following are used:

Commercialization: Commercialization comprises the set of activities in the transfer of goods and services from the point of initial production until they reach the end consumer.

Trade: the process of buying and selling or exchanging goods of different kinds.

Market: should be understood as the place where the forces of supply and demand operate, through sellers and buyers, in such a way that the transfer of ownership of goods occurs through buying and selling operations.

According to Guzman (2015), there are various methods involved in marketing that facilitate this activity, among which the following can be highlighted:

- Straight marketing;
- Indirect marketing;
- Wholesale and Retail

2.9 Types of marketing

According to Garófalo and Carvalho (2014), internal consumption, or micro-commercialization, consists of getting goods and services from the producer to the consumer or from the market to the consumer; and this distribution can be direct or indirect so that the clientele buys them, i.e. decides to sell products or services to the end user.

Micro-marketing is the realization of those activities that try to achieve the objectives of an organization by anticipating the needs of the customer and guiding a flow of goods and services that satisfy the needs of the producer to the customer.

The same author considers that external consumption or macro-commercialization can therefore be said to be an export, i.e. sending goods and services to another part of the world for commercial purposes. This can be done by different means of transport, whether land, sea or air.

2.10 Marketing channel

The marketing channel is the path taken by goods from the producer to the end consumer. It is the sequence of markets through which the product passes, under the action of various intermediaries, until it reaches the region of consumption. The marketing channel shows how intermediaries are organized and grouped together to carry out the transfer from production to consumption. The character of the various intermediaries and agencies that carry out the services of marketing a product is the arrangement and organization of the market mechanism (Frank, 2011).

The same author considers that, in this method, the human element is given special emphasis. Intermediaries are individuals or commercial organizations that specialize in carrying out the various marketing functions related to buying and selling activities, as goods move from producers to consumers. Intermediaries of direct interest to the marketing of foodstuffs can be classified as follows:

a) intermediary traders: wholesalers, retailers and speculators;

b) intermediary agents: brokers and commission agents;

c) auxiliary or instrumental organizations;

d) manufacturing industry.

Merchant middlemen hold title to the goods and therefore own the products they handle. They trade for their own profit, guaranteeing their income from the margin between the purchase and sale prices of the goods they trade. Wholesalers sell to retailers and to other wholesalers and industrialists, but do not sell significant quantities to the ultimate consumer.

Retailers buy products from wholesalers to resell to the last consumer. They are the largest group among the marketing agencies. Agent intermediaries, as they are called, act only as representatives of their clients. They don't have title and therefore don't own the goods they sell. Their income is represented by fees and commissions on the volume of sales they make (Senaes, 2017).

According to Figueiredo *et al.* (2004), commission agents or inspectors generally have a great deal of authority over the goods, and are responsible for handling them, arranging the terms of sale and deducting fees. Brokers do not regularly have physical control of the products they handle, but follow their clients' orders closely. Their negotiating powers are less than those of commission agents.

The speculative intermediaries are a group who take ownership of products in order to make a profit from short-term price fluctuations. The buying and selling activity is often carried out at market channel level. In competition with other intermediaries, these agents contribute to maintaining an appropriate price structure (Sabourin, 2010).

According to Ledesma (2017), the classification of marketing channels is based on their length and complexity. The most common types are:

- The producer sells directly to the consumer, an example of which is what happens with market traders, who are producers who sell their produce directly to the consumer.
- The operations are carried out by intermediaries. In this case, the marketing channel can vary in complexity, depending on the number of operations and therefore the number of people involved. As the economy develops and the specialization of the activity intensifies, the channel tends to become more complex.

The relationship between supply and consumer demand is widely considered to be the main determinant of market prices. However, the above analysis of market levels and marketing agents has shown that consumers and producers are separated by many intermediaries who make it possible to transfer agricultural production to final consumers. This intermediation results in marketing costs that will be higher or lower depending on the levels of intermediation (Rivas, 2007).

2.11 Product. Concept

A product is a set of attributes that the consumer considers a certain good to have in order to satisfy their needs or desires. According to a manufacturer, a product is a set of physical and chemical elements put together in such a way as to offer the user possibilities of use. It is that desired or undesired thing that a person or company makes in an exchange; the basic goal of purchasing decisions is to receive the tangible or intangible benefits associated with a product; tangible and intangible aspects include service, the retailer's image, the manufacturer's reputation and the social status associated with a product, a company's product offering is the crucial element in any marketing mix (Neves *et al.*, 2013).

Consumers are increasingly interested in buying agricultural products that are produced locally, i.e. in their region of residence or country. According to the perception of quality, the demand for fresh products, the short distance between products and consumers are some of the key factors for the demand for local products and consumers are more likely to pay a premium price for products (Singer, 2016).

2.12 Promotion

According to Kaplan and Cooper (2018), it is defined as one of the fundamental marketing tools with which the company intends to convey the qualities of its product to its customers, so that they are encouraged to purchase it; therefore, it consists of a mechanism for transmitting information. The promotion of a product is the set of activities involved in communicating the benefits of the product and persuading the market to buy from those who offer it.

2.13 Marketing agricultural products

According to Kotler (2000), agricultural marketing has become increasingly important due to:

- Increased consumer power;
- Industrialization of agribusiness;
- Increasing regulation of agricultural production;
- Progressive globalization of agriculture;
- Marketing;
- A personal attitude and a company culture.

2.14 Factors that influence the marketing process in the country.

The coordination and integration of production, storage and conservation (concentration of supply, in terms of quantity and quality), eventual industrial processing, marketing and distribution of agricultural products. Recognized by all as one of the most important limitations in the full development of Angolan agriculture, since it is accepted that its correct development will immediately trigger an incentive in all forms of agriculture which, due to the lack of the same, has often chosen not to produce, so as not to have to lose production due to the lack of an outlet (MINADER- FAO, 2003).

This marketing must go both ways if it is to be truly effective, because it must not only be able to bring inputs and other elements to the farmer, but also have the capacity to sell the products produced in the region where they exist. The prices at which the inputs are sold and the products bought must be given due consideration so as not to hinder the success and development of more remote regions, which will naturally also be linked to the ease with which the products in question can be transported by road or rail (Barros, 2007).

In this way, it can be seen that farming has problems "inside" and "outside" the farm gate, but it is undoubtedly beyond the reach of rural producers that most of the problems that affect economic and financial results occur, with adverse social consequences (Rivas, 2007).

The identification of new products that can help diversify and create added value in the domestic or foreign market will also be very important to get out of this situation, as the sector is raising awareness and mobilizing private entrepreneurs to engage in national seed production, especially of food crops such as maize, beans, rice and cassava (Mamaot, 2013).

The country has to make an effort to resolve the preconditions for the development of agriculture. It needs to create partnerships with companies with experience in the sector and take advantage of the knowledge of countries in the region (Zimbabwe, South Africa and Zambia) that produce the seeds they consume. The country invests in foreign currency to buy seeds that are not even enough to meet 50% of the country's needs in the sector. In addition to seed production, it is necessary to think about fertilizers, pesticides, tractor assembly and irrigation systems (CGA, 2010).

2.15 Market information

According to MINADER, (2020) the market information function concerns the collection, interpretation and dissemination of data in order to facilitate "marketing". An important characteristic of information is that it must be current and reliable. There are three types of information:

a) purely informative or news;

b) market analysis (perspectives) e,

c) advertising.

The "informative" type contains only data on prices, supply conditions, stock volumes, weather, among other things, without any analysis or comments on the market situation.

The "analytical" type goes beyond the news, because it provides explanations (reasons) for the current trend and makes predictions about this trend. In this case, in addition to data on the relevant variables, this data needs to be analyzed using statistical and economic models. In this case, there is a need for knowledge of factors linked to agricultural demand and supply.

The same source goes on to say that the relevant variables on the demand side include the following indicators: domestic population, level of disposable income, level of employment, per capita consumption, changes in tastes and preferences, prices of substitute goods, external demand and special government programs. On the supply side, there are: planting intentions, price expectations, prices of competitive products, expected productivity, area available for planting and adoption of technological packages.

Another type of market information can be obtained through advertising, which takes two forms. The first is called "generic", and can be conducted by the government or by a group of firms with the aim of increasing consumption of a product, without a specified brand.

2.16 Market research

In a purely business context, research into changes in consumer preferences is important for determining company policy. Thus, research into packaging in terms of shape, size, coloring, consumer behavior, sales forecasts in a certain region, research aimed at reducing "marketing" costs, the best means of communication for advertising, among others, is useful information for the company's long-term success (Junta Provincial de

Povoamento, 2015).

From this point of view, Diniz (2018) informs us that more than rural economics, marketing research is important in the following areas:

a) Demand and expenditure studies;

b) Supply studies;

c) Analysis of marketing costs;

d) Analysis of marketing margins;

e) Analysis of agricultural prices;

f) Market structure studies

III. MATERIALS AND METHODS

3.1 Location of work

In order to propose a group of actions to improve the bean marketing process, this work was carried out in the informal markets of Oshomokuyo and Germany, between February and May 2022. The province lies between latitudes 15° 10' 00" to the North, 17° 24' 00" to the South and longitudes 13° 02' 00" to the West and 17° 23' 00" to the East (Ministry of Territorial Administration, 2019).

3.2 Type and method of research

According to the nature and characteristics of this study, it was considered to be exploratory-descriptive because its characteristics make it possible to describe a population, learn about consumer behavior and qualities in relation to products, and learn about supply and demand trends.

As an empirical research method, a non-experimental cross-sectional design was used, with the use of scientific observation and research, to capture information at a single time, with the aim of describing variables and analyzing their incidence and interrelationship at a given moment, with criteria provided by Hernández-Sampieri *et al.* and Tamayo (2003).

3.3 Methodology used

The methodology used was, according to Silva (2010), a survey, defined as an in-depth research technique in which information is obtained through primary data collection, making it a qualitative study in terms of data collection and a quantitative study in terms of data interpretation. The first step was to provide a theoretical basis for the propositions put forward in this study by reviewing the literature on the subject. The information gathered from the literature review provided input for the construction of a questionnaire used to collect the data.

The questionnaire developed contained open and closed questions, which served as a guide for conducting interviews with rural producers and informal market traders. A field survey was then carried out in order to gather information on the operational structure, the flow of information and the marketing systems used by producers in the locality and in the surrounding municipalities who bring in the goods.

With regard to data collection in the field, 8 randomly selected consumers (customers) and 11 vendors were interviewed, 6 in Germany and 5 in Oshomukuyo. As a strategy for selecting the farmers and bean traders to be interviewed, two marketing outlets with a wide regional presence (two informal markets) were chosen which made the products available.

3.4 Design and sampling

The type of sampling used was simple random. The sample size was calculated using a population of 8 producers and 11 sellers, with a margin of error of 5% and a reliability of 95% in standard units corresponding to (1.96). The formula mentioned by Piza and Welsh (1968) was used.

$$n = \frac{Z^2 pq\, N}{0.05^2\,(N-1) + Z^2 pq}$$

Where:

n = sample size to be determined.

N = size of the universe (total number of elements in the population).

p = probability of the event occurring (50%).

q = probability that the event will not occur (50%).

E= maximum plausible error depending on the level of reliability desired (said error is squared)

E= 5% (0.05).

Z = Confidence coefficient for a given probability level. This coefficient squared = 95% corresponds to a value of Z= 1.96.

$$n = \frac{(1.96)\ 20.5 * 0.5 * 16}{0.052\ (16 - 1) + 1.96\ 2\ (0.5 * 0.5)}$$

n= 33,65

n= 34 slopes approximately.

3.4 Methodological steps

In order to avoid errors in the research process, different methodological stages were planned (see figure 1).

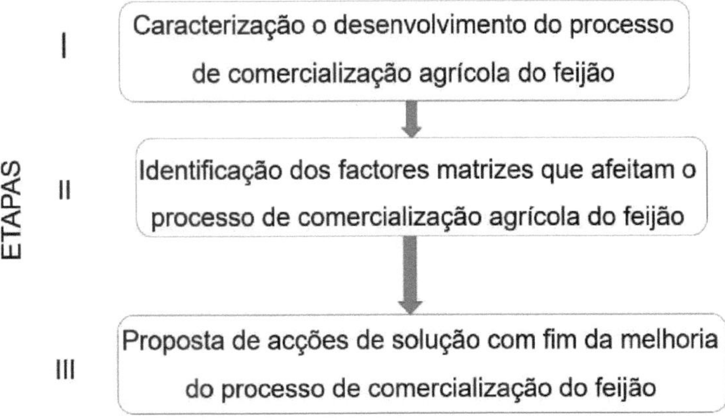

Figure 1 - Diagram of the research stages

3.5 Characterization of the development of the agricultural marketing process for beans *(Phaseolus vulgaris* L.) in the informal markets of Ondjiva Commune in Cuanhama Municipality.

To characterize the structure of the market, the methodology of Eterno and Elenor (2015) was used to represent how the bean marketing process is carried out. For this part, a diagram was made explaining the aspects that most influence the marketing process in the locality.

For this variable of establishments selling beans, the different bean sellers in the two markets were counted and the results plotted on a bar chart.

In the case of the supply-demand shift for beans in the local markets, a survey was carried out in which buyers' responses were collected and the answers represented in a table to show how the different variations occur according to sales levels and the presence of beans.

3.6 Identifying the factors that affect the process of agricultural marketing of beans *(Phaseolus vulgaris* L.) in the informal markets of Ondjiva Commune in Cuanhama Municipality.

3.7.1 SWOT analysis variables

Strengths. Maintains a high level of performance, generating present and clear advantages or benefits, with attractive possibilities in the future.

Weaknesses. It means a deficiency or lack, something in which the organization has low levels of performance and is therefore vulnerable.

Opportunities. These are those circumstances in the environment that are potentially favourable for the organization and can be changes or trends that are detected and which can be used to advantage to achieve or surpass objectives.

Threats. These are factors in the environment that result in adverse circumstances that jeopardize the achievement of the established objectives. They can be changes or trends that present themselves suddenly or gradually.

SWOT analysis

In this study, it was carried out according to the criteria of Rodríguez (2000) and Ramírez (2007), who stated that it consists of carrying out an assessment of the strong and weak factors that together diagnose the internal situation, it is a tool that can be considered easy and makes it possible to obtain an overview of the strategic situation of a given organization.

3.7 Proposed solutions to improve the process of marketing beans *(Phaseolus vulgaris* L.) in the informal markets of Ondjiva commune in Cuanhama municipality.

The training methodology described in this document is participatory. It suggests the development of group sessions, reflections, presentation dynamics, practical work and evaluation, in order to energize and strengthen the learning process and encourage new attitudes and values such as: initiative, creativity and discipline in producers.

It will also be structured in such a way as to guide the facilitator in a simple way to achieve the objectives of the workshops. The topics covered in this document include a methodology sheet with the following information:

- ❖ Objectives
- ❖ Expected goal
- ❖ Method of controlling activity

- Those responsible or involved

3.8 Organization and statistical processing of information

The information collected was carried out by coding the survey questions, drawing up figures, tables, diagrams and data matrices.

The variables that made up the data matrices are shown in the results and discussion. The research results were processed using tables and graphs. The information was analyzed using Excel for Windows version 10.

IV. RESULTS AND DISCUSSION

4.1 Characterization of the development of the agricultural marketing process for beans *(Phaseolus vulgaris* L.) in the informal markets of the Ondjiva Commune in the Cuanhama Municipality.

4.1.1 Market structure

According to the surveys carried out, which show a representation of the process, figure 2 shows the marketing channel identified for beans in the locality.

The marketing chain works under the sales model: producer-acquirer or intermediary-wholesaler and consumer. The intermediary buys the product in the production areas, on the property or in the nearest regional markets, determining the price based on supply and the expectation of the price to be obtained at the wholesale centers.

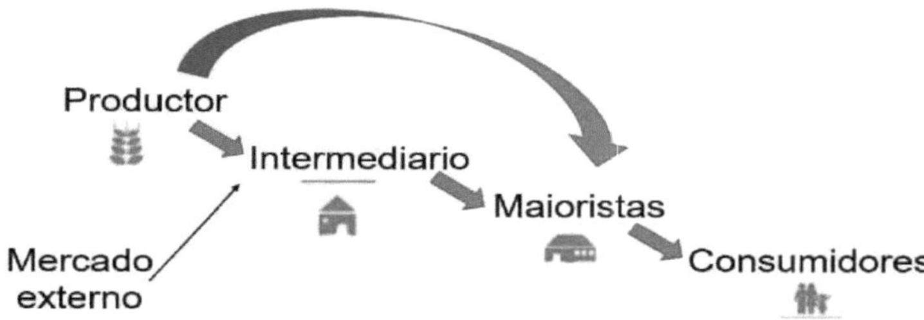

Figure 2 - Interpretation of the bean distribution chain

The market is known as the space or context where the exchange, sale and purchase of beans and services takes place between a buyer who demands them and a seller who offers them. Informal traders in the city belong to a competitive market, sometimes called a perfectly competitive market, where there are many buyers and sellers and where the goods offered by the various sellers are basically the same. Competitive buyers and sellers must accept the price that the market determines and are therefore said to be price decision-makers.

According to the analysis, a direct marketing channel is identified, in which producers are involved in selling different types of beans to the final consumer in the marketplaces or informal markets. Both buyers and sellers (in this case the producer) can be individuals and their families, agricultural companies and cooperatives (various caponos), wholesale and retail companies, companies from other sectors of the economy, service providers and governments. The end consumer can also be involved in the sale. In this respect, Rocha (2010) said that marketing should facilitate and answer the economic problems of "what" and "how much" to produce, "when", "how" and "where" to distribute the products, and in what "form".

According to this study, a lot of beans come from Huila province, which is an aspect to analyze from a market point of view. In relation to this situation, SWANDER (2014) states that the extent of the market depends on the dispersion of its consumers. However, in order for two regions (Lubango-Cunene) to be integrated into a single market, there must be the possibility of communication so that potential buyers and sellers can maintain contact and transfer ownership of the goods. For many products, however, the incorporation of several regions into the same market is limited by transportation costs. This results from the fact that trade between regions will only take place if local prices in the different regions differ by an amount greater than the cost of transportation. Otherwise, it won't pay for sellers to place their goods in the buying region.

The distribution system is made up of all the functions and actions that define the relationship between producers and their intermediaries. Its greater or lesser success depends on the level of integration and cooperation between the parties involved (SILVA, 2010). Therefore, Kotler and Armstrong (1998) consider that the greater or lesser success of the distribution system basically depends on decisions regarding channel alternatives and the level of integration and cooperation between the parties involved. Thus, they define a distribution channel as "a set of interdependent organizations involved in the process of making the product or service available to the end consumer".

4.1.2 Establishments selling beans

When we analyzed the number of bean sellers in the two markets, we found that there is a difference between the types or varieties they sell, which is greater in the Oshomokuyo informal market (see figure 3).

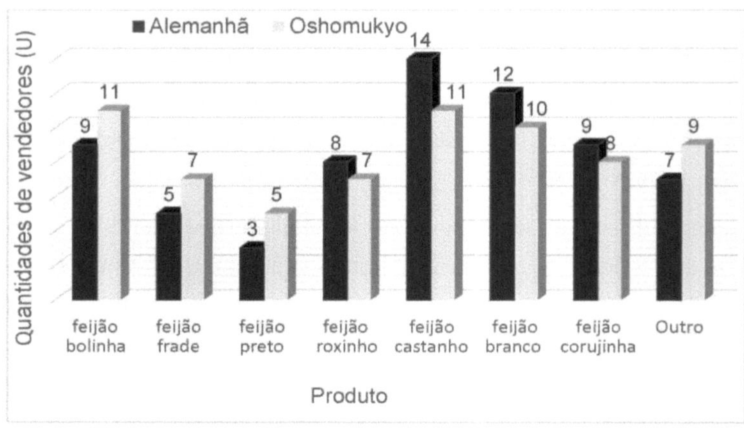

Figure 3: The presence of beans in informal markets

There is a wide range of the product on offer, but the interviewees commented that the beans most bought by customers/consumers in Germany are brown beans and white beans, with black beans and kidney beans being the least bought. This result may be due to consumption patterns or a cultural aspect of the locality. This result coincides with Singh (2017), who states that brown beans are a staple dietary component of rural and urban populations in the regions and are therefore highly consumed

In the informal market of Oshomokuyo, the results were almost similar, but it turned out that the supply of beans was greater than in the other market studied. It was difficult in the investigation process to count the quantities of beans that each vendor had and all the genotypes that are sold because of the large quantities. Despite the differences in supply in the two markets, it turned out that the prices in both were the same.

With regard to the low presence of black beans in the markets, Silva and Oliveira (2013) comment that some types of bean are generally produced infrequently due to low consumption and acceptance, and are distributed at various points of sale, but that their presence in the market is not very abundant.

The price of beans for 100% of buyers is seasonal in the second half of the year. In the months of January to April, the price is better, which may initially indicate that selling beans immediately may be the best choice for producers who can't always wait for market signals.

In this regard, Andrade *et al.* (2010) state that threshed beans, i.e. in bulk, are the most attractive to consumers and are an important purchasing option.

4.1.3 Shifting supply-demand for beans in local markets

According to the interviews carried out (see Table 1) at the Germany and Oshomokuyo markets, 92 % of the consumers surveyed said that the beans on offer are good, but they are not sold at different prices because all the vendors agree to unify the price throughout the market, which means that sales levels don't vary much.

Table 1. Analysis of the effects of the simultaneous displacement of demand and supply in the two local markets.

Displacement	Increased demand	Constant demand	Reducing demand
Increasing supply	Price? Quantity ↑	Price ↓ Quantity ↑	Price ↓ Quantity?
Constant supply	Price ↑ Quantity ↑	There are no changes	Price ↓ Quantity ↓
Reducing supply	Price ↑ Quantity ?	Price ↑ Quantity ↓	Price? Quantity ↓

It's important to note that in the real world there are simultaneous changes in the demand and supply curves in both markets. Therefore, an increase in supply may not be accompanied by a reduction in prices, because there may also have been an increase in demand. Table 1 above shows the possible effects of shifting supply and demand on the market equilibrium price and quantity

The price of the good or service is the main information available for decision-making, with buyers wanting to pay the lowest possible price and sellers wanting to charge the highest possible price, especially for the Lebanese in the city of Ondjiva who were not part of this study. In general, chaos does not occur, and there are relatively long periods of time when prices are stable, when they oscillate around a stable value, called market equilibrium (ME) in economic theory.

Thus, the model proposes that the quantities demanded and supplied will adjust to a price level that meets the objectives of buyers and sellers (Hall and Lieberman, 2013; Pinho *etal.* 2014).

The main variable is the price of the good or service. The supply model predicts that when the price of a good rises and all other variables remain unchanged, the quantity offered of that good increases. This is because the higher price increases profitability, making sellers want to increase their supply of beans.

4.2 Identifying the main problems affecting the marketing of beans in Cunene's

informal markets.

4.2.1 SWOT matrix

Taking into account the internal and external diagnosis shown in Table 2, the administrative tool Matrix (SWOT) was used for beans in the two informal markets studied in order to identify strategies.

The strengths, opportunities, weaknesses and threats included in the diagnosis are specific to the trade process here in the province, so these variables are unique in their operation and results and should not be generalized.

Table 2. Strengths, opportunities, weaknesses and threats

Forces	A traditional product, much appreciated throughout the province; A hardy and resistant crop, it is an important source of energy with a low fat content; It is adaptable to different types of climate and soil and can be grown on its own, in a consortium or intercropped; Research, funding and production support bodies promote innovation in the production chain, overcoming challenges related to new pests, increasing productivity and the necessary investments; The existence of three harvests makes it easier to change planting intentions throughout the year, which can influence prices.
Weaknesses	Product with short staking time; Concentration of production on beans, which are well accepted domestically, but produced little for export; Family production is low-tech and professional, using home-grown seeds, which degenerates the varieties planted and facilitates contamination by pathogens and mechanical damage, as a result of the low capitalization of producers and poor technical assistance
Opportunities	Research into new ready or semi-ready products with beans, reducing preparation time, such as hamburgers, flour (gluten-free) for making bread, cookies and pasta; Research into new varieties that are earlier, more productive or have larger grains (European preference) to avoid concentrating on beans and offer more options on the market; Development of traceability systems, due to the changing profile of consumers, who are more demanding and seek to know the origin and conditions in which the product is grown; Marketing campaigns promoting the nutritional qualities of beans.

Threats	Climate change tends to make extreme events such as droughts or floods more severe, more intense and with shorter cycles of occurrence. The emergence of new pests and diseases that are resistant to pesticides.

The situations encountered in the two places presuppose similar reactions, for example, a weakness or threat in the German square will not necessarily be so for Oshomokuyo.

In this sense, Steiner George (1995) states that proposals for improvement should be aimed at reducing weaknesses, reinforcing and maintaining strengths, searching for suitable opportunities for capabilities, and providing a defense against external threats. In another work, Thompson and Strickland (2001) published that the improvements accepted must be congruent with the conditions or means of operation of the marketing process, in order to increase its abilities and resources.

We found a series of forces and opportunities facing the process, which, when taken advantage of, could reverse the development of the bean agricultural trade in the province, with benefits for the family economy, but specifically for improving agricultural management, the availability of mechanisms and sources for acquiring inputs for improving infrastructure and, in the long term, improving prices, as well as the availability of beans in the markets. A group of threats have been identified which, over time, imply a regression in this field, which can be reversed by achieving unisectorality, so that the institutions involved and present in the province can work towards a common goal.

4.3 Proposed actions for the process of marketing beans *(Phaseolus vulgaris* L.) in the informal markets of Ondjiva Commune in Cuanhama Municipality

4.3.1 Proposed actions

The above actions must be carried out in different ways, but the most efficient way is through government action projects with a clear intention to implement them and put them into practice, or through agricultural extension, an aspect that has been little explored in the province and is also limited (see table 3).

Related to the above, the identification of weaknesses is the first step towards improving the living conditions of the population through defined, decisive and concerted action between local and public agents, based on the use of business management capacities and the creation of an innovative territorial environment to implement the proposed actions.

Table 3. Proposed actions for the bean marketing process *(Phaseolus vulgaris* L.)

WEAKNESS FOUND	ACTION	FUTURE RESULTS	MONITORING INDICATORS		
			Time	**Responsible**	**Control**
Product with short staking time.	Ensuring the storage of volumes of beans during harvest periods.	Increased staking period; Increased presence of beans at times of low production.	Once a year;	Salesman; Peasants.	Chequeo, Periodic review by MINAGRI experts
Objective:	Increasing the storage capacity of beans.				
Concentration of bean production, well accepted domestically, but little produced for export.	Draw up a marketing strategy that provides for the distribution of beans according to the different times of the year.	Achieving an equitable presence of commercial quantities of beans throughout the year	annually	IDA, MINAGRI	check, Periodic review by MINAGRI experts
Objective:	Increase customer satisfaction by increasing the presence of beans on the market throughout the year.				
Inadequate hygiene conditions	Using packaging and improving the presentation of beans in markets	Maintaining hygienic sanitary conditions	Twice a year.	Sales, MINAGRI	check,
	to preserve its harmlessness.	when the beans are sold in the markets.			Periodic review by MINAGRI experts
Objective:	Improving the quality and hygienic conditions of beans sold on the market.				

Family production is low-tech and professionalized, using home-grown seeds, which degenerates the varieties planted and facilitates contamination by peptogens and mechanical damage, as a result of poor technical assistance.	Design development projects to introduce new technologies. Design and manage a training program	Increasing the use of technology and professionalizing technical assistance	Twice a year	MINAGRI, extension workers, IPO.	Report and reports.
Objective:	Increasing technology in bean production and professionalization with a view to improving the crop in the locality.				
High number of competitors	Competing in a market where there are other bean sellers.	Maintain stable earnings.	Weekly	Salesman; Peasant.	check, Periodic review by experts.
	Offer beans at different prices.	Increase sales levels.			
Objective:	Guaranteeing bean sales levels, although there is a high level of competence.				

V. CONCLUSIONS

a) Beans are marketed using the sales model: producer-acopiator or intermediary-wholesaler and consumer.

b) At the markets in Germany and Oshomokuyo, 92 % of the consumers surveyed said that the beans on offer are good, but they are not sold at different prices, and at Oshomokuyo, there is a greater variety of beans on sale compared to Germany.

c) The situations found in the SWOT in the two places presuppose similar reactions, but it has been proven that a weakness or threat in Germany will not necessarily be for the Oshomokyo market.

d) The above actions must be carried out in different ways, but the most efficient way is through government action projects with a clear intention in their execution.

VI. RECOMMENDATIONS

a) Deepen the study of the bean marketing process.

b) Implement the proposed actions set out in the paper.

c) Socialize this work in different scenarios.

VII. BIBLIOGRAPHICAL REFERENCES

- ✓ Agronégocio. (2014). *Cooperation networks in the Agribusiness technologies and services sector.* Luanda.
- ✓ Almeida, A. L. G. DE, Alcântara, R. M. C. M. DE Nobrega, R. S. A., Nobrega, J. C. A., Leite, L. C., Silva, J. A. L. (2010). *Productivity of cowpea cv BR 17 Gurguéia inoculated with symbiotic diazotrophic bacteria in Piauí.* Revista Brasileira de Ciências Agrárias, v. 5, n. 3, p. 364 - 369.
- ✓ Andrade, F.N., Rocha, M. de M., Gomes, R.L.F., Freire Filho, F.R., Ramos, S.R.R. (2010) *Estimates of genetic parameters in cowpea genotypes evaluated for fresh beans.* Revista Ciência Agronômica, Fortaleza, v. 41, n. 2, p. 253-258, abr./jun.
- ✓ Arbage, A. P. (2016). *Fundamentals of Rural Economics.* Chapecó: Argos, 20pp.
- ✓ Barros, G. (2007). *Economics of agricultural marketing.* Center for Advanced Studies in Applied Economics - CEPEA. University of São Paulo - USP.
- ✓ Barros, G.S.A.C., J.G.Martines Filho, (2017). *"Price Transmission and Marketing Margins for Agricultural Products" in DELGADO,* G.C., J.G. GASQUES and C.M. VILLA VERDE (org.), Agriculture and Public Policies. IPEA Series no. 127, Brasília- DF
- ✓ C.G.A. (2010) *Consulate General of Angola.* Kwanza Norte, Capital: N dalatando. Accessed on May 15, 2021, on the website of: Consulate General of Angola: http://www.consuladodeangola.org/index.php?Itemid=168&id=188&option=co m_c ontent&tas k=view
- ✓ Castiñeiras L. (2011). *Management and in situ conservation of genetic resources of plants cultivated in homestead gardens in Cuba.* Organic Agriculture. 1. La Habana. Cuba
- ✓ DIARIO DA REPUBLICA (2014). *National Strategy for Rural Trade and Entrepreneurship.* Official Gazette, 2
- ✓ Diniz, A. C. (2018). *Angola's Physical Environment and Agricultural Potential.* Portuguese Cooperation Institute. Lisbon.
- ✓ Eterno, P. V. and Elenor, A. A. W. (2015). *Analysis of the Commercialization Channels for Common Beans in the Production Hubs of the Eastern Region of the*

State of Goiás. Revista Cojuntura económica goania, Jun, No:33,16-25pp.

- FAO. Food and Agriculture Organization of the United Nations. FAOSTAT. (2019). Available at: <http://faostat.fao.org/default.aspx>. Accessed on: 26 Dec. 2021.
- FAO. Food and Agriculture Organization of the United Nations. FAOSTAT. (2020). Available at: <http://faostat.fao.org/default.aspx>. Accessed on: 21 Jan. 2022.
- Ferreira, C. M., Peloso, M. J. Del, Faria, L. C. (2012). *Beans in the national economy.* Santo Antonio de Goiás: Embrapa Arroz e Feijão. 47 p. (Embrapa Arroz e Feijão. Documentos, 135).
- FIEP (Federation of Industries of the State of Paraná) (2017). *Program to increase sales of Paraná products.* FEIJÃO - Version 1.0. May 2006. Available at: <http://www.fiepr.org.
br/fiepr//conselhos/agroindustria_alimentos/uploadAddress/
Relat%C3%B3rioFeij%C3%A3o0506.pdf>. Accessed on: 01 Jan. 2021.
- Figueiredo, A., Prescott, E., Melo, M. F. de. (2004). *Integration between the family market and the retail market. Brasília:* Universa. 196p.
- Fisher, L. and Navarro, A. (2018). *Introducción a la Investigación de Mercados,* Ed. McGraw Hill, 2ª Ed, México, 162 pp.
- Frank, H. F. (2011). *Microeconomics and behavior.* São Paulo: McGraw-Hil. 3rd ed. p. 300-379.
- Garófalo, G. de L; Carvalho, L. C. P. de. (2014). *Theoretical microeconomics.* São Paulo: Atlas S.A. 2nd ed. p. 170-260.
- Gepts, P., Debouck, D. G. (2015). *Origin, domestication and evolution of the common bean (Phaseolus vulgaris L.).* In: SCHOONHOVEN, A. van; VOYSEST, O. Common bean research for crop improvement. Wallingford: CAB International; Cali: CIAT. p. 7-53
- González M, (2018). *Fungal diseases of the Fríjol.* Editorial Científica técnica. La Habana. Cuba
- Guzmán, E. S. (2015). *Agroecology and sustainable rural development.* In: AQUINO, A. M.; ASSIS, R. L. Agroecology: principles and techniques for sustainable organic agriculture. Brasília, DF: Embrapa. p. 101-131.

✓ Hall, M., Lieberman, M. (2013). *Introduction to futures and options markets.* São Paulo, Bolsa de Mercadorias & Futuros (BM&F) - Cultura Editores Associados, 2nd expanded edition, 19pp.

✓ Hernández, S.R., Colinas, C. and Baptista, P. (2003). *Research Methodology,* 684pp.

✓ Joseph, E. (2015). *The guru's guide: the best concepts in business practices.* São Paulo: Campus. p 189-204.

✓ Journal de Angola (2021). *Bean production in deficit by 350,000 tons.* News article. Available: https://www.jornaldeangola.ao/ao/noticias/producao-de-feijao-com-defice-de-350- thousand-tonnes/ consulted:11/07/2022

✓ Junta Provincial de Povoamento (2015). *Application of the method for planning an agricultural enterprise located in the village of Luinga*, municipality of Ambaca, district of Kwanza Norte, (Angola). Reordenamento, pp. 3-7.

J Kaplan, L. (2016). *ArchaeologicalPhaseolusfrom Tehuacán.* In: BEYERS, D. (Ed.). The prehistory of the Tehuacán Valley: environment and subsistence. Austin: University of Texas. v. 1, p. 201-212

✓ Kaplan, R. S., Cooper, R. (2018). *Cost and performance: manage your costs to be more competitive.* São Paulo: Futura. 366 p.

✓ Kotler, Philip (2000). *Principles of marketing.* São Paulo: Prentice Hall.

✓ Ledesma, A. M. (2017). *Agronegocios, empresa y empreendimento.* 2ª ed. Buenos Aires: El Ateneo. 266 p.

✓ Mamaot (2013). *Strategy for the Valorization of Local Agricultural Production.* Final Report of the GEVPAL Working Group. Ministry of Agriculture, Sea, Environment and Spatial Planning. Angola

✓ Mendes, Judas Tadeu Grassi; PADILHA JUNIOR, João Batista (2007). *Agribusiness: an economic approach.* São Paulo: Pearson Prentice Hall,

✓ MINADER- FAO. (2003). *Review of the agricultural sector and the food security strategy for defining investment priorities* (TCP/ANG/2907) ± Agricultural Production Systems. Working Document No. 07. Preliminary version for comments. Accessed on May 13, 2021, on the website of: Angolan Ministry of Agriculture and Rural Development:

http://www.minader.org/pdfs/fomento/volume_iii/sistema_producao_agricola.pdf

- MINADER, (2020) *Construtora Norberto Odebrecht S. A. and Sondotécnica Engenharia de Solos S. A.* Matala ± Capelongo General Conductor Canal Rehabilitation Project. Huíla Province ± Matala Municipality.
- Moreira, I. (2021). *Angola Agriculture, Natural Resources and Rural Development.* Volume I. ISA-Press. Lisbon.
- Muñoz, G. and Singh, S. (2003). *Comparative studies of sources of resistance to Bacteriosis común disponibles en diferentes especies de phaseolus y progreso genético a través de cruzamientos ínter específicos y piramidación de genes.* En S. P. Singh & O. Voysest, (Eds.), Taller de mejoramiento de frijol para el Siglo XXI: Bases para una estrategia para América Latina. CIAT. Cali. Colombia.
- Navolar, T. S., Rigon, S. do A., Philippi, J. M. de S. (2010). *Dialogue between agroecology and health promotion.* Revista Brasileira em Promoção da Saúde, Fortaleza, v. 23, n. 1, p. 69-79.
- Neves, M. F. (Org.) (2010) *Economia e gestão dos negócios agroalimentares: indústria de alimentos, indústria de insumos, produção agropecuária, distribuição.* São Paulo: pioneira. ch. 3, p. 39-60.
- Neves, M. F., Castro, L. T. (2017). *Marketing and strategy in agribusiness and food.* São Paulo: Atlas, 365 p.
- NEVES, Marcos Fava; Chaddad, Fabio Ribas; Lazzarini, Sérgio Giovanetti. (2013). *Food business management.* São Paulo: Pioneira Thomson Learning, 14pp.
- Pinho, Diva Benevides; VASCONCELLOS, Marco Antônio Sandoval de (Org.) (2014). *Handbook of Economics.* 5. ed. São Paulo: Saraiva,33pp.
- Piza, C.T., R.W. Welsh, (2018). *Introduction to Commercialization Analysis.* Handout Series No. 10. Department of Economics - ESALQ/USP, Piracicaba-SP.
- Porter, Michael E. (2006) *Competitive strategy: techniques for analyzing industries and the competition.* Rio de Janeiro: Campus, 32pp.
- Ramírez, R. J. L. (2007). *Material del curso: Gestión estratégica,* Maestría en Ciencias Administrativas, IIESCA UV, México.
- Rivas, G. (2007). *Economics of agricultural marketing.* Center for Advanced

Studies in Applied Economics - CEPEA. Agustino Neto University,23-35pp

- Rocha, M. de M. (2010) *The cowpea for consumption as fresh grain.* Agrosoft Brasil, 11pp. Available at: http://www.agrosoft.org.br/agropag/212374.htm. Accessed on: July 5, 2021.
- Rocha, M.M., Freire-Filho, F.R., Ribeiro, V.Q., Carvalho, H.W.L., Belarmino-Filho, J., Raposo, J.A.A., Alcântara, J.P., Ramos, S.R.R., Machado, C.F. (2007). *Adaptability and productive stability of semi-upright cowpea genotypes in the Northeast Region of Brazil.* Pesquisa Agropecuária Brasileira, 42: 12831289.
- Rodríguez, V. J. (2000). *Management with a Strategic Approach.* Edit. Trillas, Mexico.
- Sabourin, E. (2010). *Multifunctionality and non-market relations: management of common resources in the Northeast.* Caderno do C R H, n. 22, v. 56, p. 151-169.
- Scubla L. (1985). *Logiques de la réciprocité.* Paris, Ecole Polytechnique, Cahiers du CREA n. 6, 283p.
- SENAES, National Secretariat for Solidarity Economy (2013). *Solidarity Economy Program in Development, Brasília:* SENAES-MTD.
- Senaes. (2017). *Fair and solidarity trade bill (CJS).* Brasília: SENAES-MTD
- Silva, E. F. (2018). *Growing beans - escala rural magazine.* São Paulo: Escala, n. 5, p. 10 - 17.
- Silva, I. (2010). *The commercialization process in the company fazenda sapucaia, municipality of santa isabel do pará.* Administration & Technology, v. 1, n. 1.
- Silva, P.S.L., Oliveira, C.N. (2013). *Yields of green and mature beans from caupi cultivars.* Horticultura Brasileira, Brasília, v. 11, n. 2, p. 133-135.
- Singer, P. (2016). *Introduction to Solidarity Economy.* São Paulo: Perseu Abramo, 127p.
- Singh, B.B. (2017). *Recent progress in cowpea genetics and breeding.* Acta Horticultura, The Hague, n. 752, p. 69-76, 2007. Edition of the Proceedings of the International Conference on Indigenous Vegetables and Legumes, Hyderabad, India, Sep. Available at: http://www.actahort.org/books/752/752 7.htm
- Socorro, M and Martín D. (2009): *Granos, Editorial Pueblo y Educación.* La Habana. Cuba
- Steiner-George, A. (1995). *Strategic Planning.* Edit. CECSA, Mexico.

- Tamayo, M. (2003). *Elproceso de investigación científica,* 440 pp., Limusa-Noriega Editores. México, 43pp.
- Thompson, L. and Strickland, A. (2001). *Administración Estratégica.* Edit. MacGraw-Hill, Colombia.
- Unifeijão (2011). *Bean classification.* Available at: <http://www. unifeijao.com.br/telas/class_feija.php>. Accessed on: 29 Dec. 2021.
- Wander, A. E. (2007). *Bean production and consumption in Brazil, 1975-2005.* Informações Econômicas, São Paulo, v. 37, n. 2, feb.
- Wander, A.E. (2014). *Irrigated bean cultivation in the northwest region of Minas Gerais.* Santo Antônio de Goiás: Embrapa Rice and Beans. Available at: <http://sistemasdeproducao.cnptia.embrapa.br/FontesHTML/Feijao/FeijaoIrriga do NoroesteMG>. Accessed on: Jan. 2021

VIII. ANNEXES

Annex 1.

Bean genotypes on the market in Germany

Oshomukyo survey day

Inquiry day in Germany

ANNEX 2

Name of respondent:

Age:.__ Place:. ______________________________ Data:. _________________

Main channels used by copiers to market beans

() Associations

() Fees

() Intermediaries

() Supermarkets, stores

() Direct sale at the place of production

() Other:. ____

Main varieties you sell?

In what unit of measurement is the product sold?

In kilograms ________ in pounds________ tons ________ other ________

What would help you sell more?

Who buys more products to consume?

Women __________ men __________

When do prices rise?

How do you trade beans?

How many kg do you sell in a day?

Which beans are bought the most and the least?

Printed by Books on Demand GmbH, Norderstedt / Germany